THE POETRY OF DYSPROSIUM

The Poetry of Dysprosium

Walter the Educator

Silent King Books a WhichHead Imprint

Copyright © 2024 by Walter the Educator

All rights reserved. No part of this book may be reproduced in any manner whatsoever without written permission except in the case of brief quotations embodied in critical articles and reviews.

First Printing, 2024

Disclaimer
This book is a literary work; poems are not about specific persons, locations, situations, and/or circumstances unless mentioned in a historical context. This book is for entertainment and informational purposes only. The author and publisher offer this information without warranties expressed or implied. No matter the grounds, neither the author nor the publisher will be accountable for any losses, injuries, or other damages caused by the reader's use of this book. The use of this book acknowledges an understanding and acceptance of this disclaimer.

"Earning a degree in chemistry changed my life!"
- Walter the Educator

dedicated to all the chemistry lovers, like myself, across the world

CONTENTS

Dedication v

Why I Created This Book? 1

One - Oh, Dysprosium 2

Two - Symbol Of Wonder 4

Three - Glorious Sight 6

Four - Element Rare 8

Five - Dysprosium, A Gift 10

Six - Treasure Of The Periodic Table 12

Seven - Seamlessly Embrace 14

Eight - Strength And Might 16

Nine - Conductor Of Power 18

Ten - Testament To Science 20

Eleven - Dysprosium, A Puzzle 22

Twelve - Luminescent Glow 24

Thirteen - Renewable Energy	. . . .	26
Fourteen - Hidden Gem	. . . .	28
Fifteen - Glorious Spark	. . . .	30
Sixteen - Discovery And Exploration	. . .	32
Seventeen - Element Extraordinary	. . . .	34
Eighteen - Atomic Wonders	. . . .	36
Nineteen - Science And Industry	. . . . .	38
Twenty - Scientific Success	. . . .	40
Twenty-One - Inspire Brilliance	. . . . . .	42
Twenty-Two - Metal Of Rare Might	. . . .	44
Twenty-Three - Captivating Grace	. . . .	46
Twenty-Four - Infinite Worth	. . . . . .	48
Twenty-Five - Sustainable Ways	. . . . .	50
Twenty-Six - Scientific Throng	. . . . . .	52
Twenty-Seven - Electric Cars	. . . . . .	54
Twenty-Eight - Element Divine	. . . . . .	56
Twenty-Nine - Firmly Stand	. . . .	58
Thirty - Fulfilling Dreams	. . . .	60
Thirty-One - Catalyst For Progress	. . . .	62
Thirty-Two - Fascination	. . . .	64

Thirty-Three - Guardian Of Stability 66

Thirty-Four - Dance Of Electrons 68

About The Author 70

WHY I CREATED THIS BOOK?

Creating a poetry book about the chemical element of Dysprosium was an intriguing and unique idea. Dysprosium is a rare earth metal with various fascinating properties, making it an interesting subject for exploration through poetry. By writing poems about Dysprosium, I can delve into its atomic structure, physical characteristics, historical significance, and potential applications. This can provide an opportunity to blend scientific knowledge with artistic expression, creating a bridge between the worlds of science and poetry. This book can educate and entertain readers, offering a fresh perspective on a lesser-known element.

ONE

OH, DYSPROSIUM

In the realm of elements, rare and true,
Lies a metal so unique, a shade of blue.
Dysprosium, its name rings through,
A chemical wonder, a mystery to pursue.

In the depths of the periodic table it resides,
A treasure hidden, where few eyes have spied.
With atomic number sixty-six, it thrives,
Its properties, a tale that nature provides.

Magnetic and strong, it holds its sway,
With its magnetic moment, it leads the way.
In magnets and lasers, its powers play,
Guiding our technology day by day.

Dysprosium, a guardian of light,
Absorbing and emitting with all its might.
In fluorescent lamps, shining ever bright,
Illuminating our world through the night.

A catalyst, it aids in chemical reactions,
Transforming compounds with precise actions.
In organic synthesis, it finds attractions,
Creating bonds and new chemical fractions.

Oh, Dysprosium, a gem in the periodic chart,
A symbol of knowledge, a work of art.
With your unique properties, you impart,
A legacy that science will never depart.

So let us admire this element rare,
And marvel at its wonders beyond compare.
Dysprosium, a testament to nature's flair,
A symbol of scientific endeavor and care.

TWO

SYMBOL OF WONDER

In the realm of elements, a secret it holds,
Dysprosium, a metal with tales untold.
With atomic number sixty-six, it's bold,
A captivating story about to unfold.

In the depths of the periodic chart,
Dysprosium stands, a work of art.
Its magnetic nature sets it apart,
A force of attraction, a mystical start.

Like a magnet, it pulls, it draws,
Aligning its electrons with cosmic laws.
Its magnetic moment, nature's applause,
A dance of forces, without a pause.

Dysprosium, a guardian of light,
In fluorescent lamps, it shines so bright.

A luminescent glow, a captivating sight,
Guiding us through the darkest night.

In the world of chemistry, it plays a role,
A catalyst, speeding up reactions so bold.
With precision and power, it takes control,
Creating new compounds, a story to unfold.

Oh, Dysprosium, a jewel of the periodic chart,
With your properties unique and rare, you impart.
A tale of mystery, a scientific art,
Forever etched in the annals of our heart.

So let us celebrate this element divine,
With its magnetic allure and radiant shine.
Dysprosium, a treasure we're fortunate to find,
A symbol of wonder in the grand design.

THREE

GLORIOUS SIGHT

In the realm of chemistry, a metal of might,
Dysprosium emerges, a captivating light.
Atomic number sixty-six, shining so bright,
A tale of fascination, an ethereal flight.
    A rare earth element, with secrets concealed,
Dysprosium's allure, yet to be revealed.
Magnetic in nature, its powers concealed,
A force of attraction, a magnetic field.
    Within its core, a magnetic moment resides,
Guiding compass needles, where destiny abides.
With its strength and power, it bravely strides,
In the realm of magnets, where it presides.
    Dysprosium, a catalyst of reactions profound,
In the world of chemistry, its influence unbound.

Facilitating change, with precision it's crowned,
Transforming compounds, with magic renowned.

    Oh, Dysprosium, a symbol of strength and grace,
In our technological world, you find your place.
In lasers and lighting, a luminous embrace,
Illuminating our lives, with your radiant trace.

    A guardian of light, shining through the night,
Dysprosium's luminescence, a glorious sight.
In the periodic table, a jewel shining bright,
A testament to science, a marvel of might.

    So let us marvel at this element rare,
Dysprosium, a wonder beyond compare.
In its mysteries, we find answers to share,
A symbol of science's unending affair.

FOUR

ELEMENT RARE

In the realm of elements, a hidden treasure lies,
Dysprosium, a metal that captivates the eyes.
Atomic number sixty-six, its secret disguise,
A symphony of properties, a splendid surprise.
   Magnetic guardian, with powers untold,
Dysprosium's allure, a magnetism so bold.
Aligning its spins, a story to unfold,
A force of attraction, a tale to be told.
   In the world of light, it casts a vibrant hue,
Dysprosium's fluorescence, a spectacle to view.
In phosphors and lasers, its brilliance shines through,
Illuminating our world, with colors anew.
   A catalyst supreme, in chemical reactions,
Dysprosium's presence, a catalyst of attractions.

Speeding up transformations, with precise actions,
Unleashing the potential of chemical transactions.

Oh, Dysprosium, a symbol of strength and grace,
A marvel of nature, in the periodic space.
With your magnetic charm and luminescent embrace,
You leave an indelible mark, a legacy in place.

So let us celebrate this element rare,
Dysprosium, a wonder beyond compare.
In its mysteries, the scientific community shares,
A testament to curiosity, a journey that dares.

FIVE

DYSPROSIUM, A GIFT

In the heart of the elements, a jewel so rare,
Dysprosium, a treasure beyond compare.
With atomic number sixty-six, it takes its stance,
A magnetic marvel, a captivating dance.

Magnetism's guardian, it holds the key,
With magnetic moments, it sets us free.
Drawing us closer, like a cosmic embrace,
Dysprosium's allure, no one can erase.

In the realm of light, it casts its spell,
Fluorescent enchantment, it knows so well.
Radiant glow, illuminating our way,
Dysprosium's brilliance, bright as day.

A catalyst of change, in the realm of reaction,
Dysprosium's touch ignites chemical passion.

Speeding up transformations, with a magical touch,
Creating new compounds, it inspires so much.

Oh, Dysprosium, a symbol of strength and might,
In the periodic table, a beacon of light.
With your unique properties, you take flight,
A testament to science, shining so bright.

Let us marvel at this element divine,
Dysprosium, a gift in nature's design.
In its mysteries, we seek and find,
A world of wonder, forever entwined.

SIX

TREASURE OF THE PERIODIC TABLE

In the depths of the periodic chart, it resides,
Dysprosium, a metal that subtly hides.
With its magnetic charm, it captivates and guides,
A symbol of strength, where science collides.

Dysprosium, a guardian of the magnetic field,
In magnets and motors, its powers revealed.
With its magnetic moments, it never yields,
Attracting and repelling, a force unconcealed.

In the realm of light, its luminescence gleams,
Dysprosium's radiance, beyond our dreams.
In phosphors and lasers, its brilliance streams,
Illuminating our world with vibrant beams.

A catalyst supreme, in the world of chemistry,
Dysprosium's presence sparks a symphony.

Accelerating reactions, with precision and glee,
Creating new compounds, a marvel to see.
    Oh, Dysprosium, a treasure of the periodic table,
With your unique properties, you enable
Technological advancements, a legacy stable,
A symbol of progress, forever unforgettable.
    Let us celebrate this element, rare and distinct,
Dysprosium, a marvel that makes us think.
In its atomic wonders, we find a link,
To the mysteries of nature, in harmonious sync.

SEVEN

SEAMLESSLY EMBRACE

Dysprosium, a whisper in the air,
A secret hidden within its atomic lair.
With magnetic allure and radiance rare,
It dances with the elements, beyond compare.

In high-tech magnets, its power resides,
Guiding the currents, where science presides.
A magnetism so strong, it defies,
Pulling us closer, as curiosity guides.

Oh, Dysprosium, an enigmatic force,
In the realm of chemistry, you chart a course.
A catalyst, sparking reactions with no remorse,
Unleashing transformations, a cosmic discourse.

In the tapestry of elements, you shine,
A symbol of knowledge, forever intertwined.

With your luminescent glow, so divine,
You illuminate our world, a celestial sign.
    Let us marvel at this element's grace,
Dysprosium, a puzzle we eagerly chase.
In its mysteries, we find a sacred space,
Where science and wonder seamlessly embrace.

EIGHT

STRENGTH AND MIGHT

In the realm of elements, a jewel so rare,
Dysprosium, with secrets beyond compare.
With atomic number sixty-six, it stands,
A marvel of nature, crafted by cosmic hands.

Dysprosium, a magnetism it claims,
Attracting attention with its magnetic flames.
In sophisticated technology, it plays a key role,
From data storage to electric motors' control.

With its luminescent hues, it captivates the eye,
A shimmering spectacle, lighting up the sky.
In phosphor screens, it paints a vivid scene,
A symphony of colors, a visual dream.

A catalyst, it speeds up chemical reactions,
Unleashing transformations with precision and

actions.
Dysprosium, a silent hero behind the scenes,
Empowering progress, fulfilling our dreams.

Oh, Dysprosium, a symbol of strength and might,
In the periodic table, shining so bright.
A testament to science's endless quest,
To unravel the universe, its secrets manifest.

Let us celebrate this element's might,
Dysprosium, a star in the scientific light.
In its atomic wonders, we find delight,
A tribute to nature's grand design, sublime and infinite.

NINE

CONDUCTOR OF POWER

In the land of elements, a gem so rare,
Dysprosium, with its secrets to share.
A magnetic force, it pulls with might,
Guiding us through the realms of light.

Oh, Dysprosium, a conductor of power,
In magnetic fields, you never cower.
With your atomic structure, so strong and sound,
You bring stability, making bonds profound.

A luminescent dance, it performs with grace,
In phosphor screens, a captivating embrace.
Colors dance and shimmer, with vibrant delight,
Dysprosium's radiance, a mesmerizing sight.

A catalyst supreme, it speeds up reactions,
Transforming compounds, igniting chemical attrac-

tions.
Unlocking the mysteries, it holds the key,
To a world of possibilities, for all to see.

 Oh, Dysprosium, a symbol of innovation,
In the realm of science, you fuel our fascination.
With your unique properties, you inspire,
Pushing boundaries, taking us higher.

 Let us marvel at this element's allure,
Dysprosium, a treasure so pure.
In its atomic wonders, we find delight,
A testament to nature's brilliance, shining bright.

TEN

TESTAMENT TO SCIENCE

In the realm of elements, a star is born,
Dysprosium, with mysteries to adorn.
A magnetic force, pulling us near,
Its allure captivating, crystal clear.

Within its atomic structure, secrets reside,
A symphony of electrons, dancing in stride.
Unraveling the complexities, we dare to explore,
Dysprosium, a marvel we can't ignore.

With luminescent hues, it paints the night,
Phosphorescent glow, a celestial light.
Guiding our way through the darkest of times,
Dysprosium, a beacon, forever sublime.

A catalyst of change, it sparks the flame,
Accelerating reactions without any shame.

From laboratory benches to industry's might,
Dysprosium, a catalyst of infinite height.

    Oh, Dysprosium, a symbol of power and grace,
In the periodic table, you find your rightful place.
With unique properties, you captivate our gaze,
Unleashing wonders, shaping future's maze.

    Let us celebrate this element divine,
Dysprosium, a treasure that will forever shine.
In its atomic wonders, we find inspiration,
A testament to science, a source of elation.

ELEVEN

DYSPROSIUM, A PUZZLE

In the realm of elements, a luminary is found,
Dysprosium, a jewel, so profound.
With magnetic allure, it pulls us near,
A symphony of electrons, a cosmic frontier.
   Oh, Dysprosium, an enigmatic force,
With atomic secrets, you chart your course.
In alloys and magnets, your power resides,
Guiding the currents, where innovation abides.
   A catalyst of change, you speed reactions along,
Creating possibilities where they belong.
In labs and industries, your magic unfolds,
Transforming the ordinary into extraordinary gold.
   Oh, Dysprosium, a symbol of strength and might,
You illuminate the path with your radiant light.

A testament to science's relentless quest,
To understand the universe, we are blessed.
   Let us marvel at this element's grace,
Dysprosium, a puzzle we eagerly embrace.
In its mysteries, we find a sacred space,
Where science and wonder dance, interlace.

TWELVE

LUMINESCENT GLOW

In the realm of elements, a gem does shine,
Dysprosium, a treasure of atomic design.
With properties rare, it captures the eye,
A marvel of science, a star in the sky.

Magnetic in nature, its power holds sway,
Guiding the fields, as they ebb and they sway.
In motors and turbines, its strength is renowned,
Efficiently turning, without making a sound.

Oh, Dysprosium, your luminescent glow,
Lights up the darkness, a celestial show.
In phosphors and lasers, your brilliance displayed,
A symphony of colors, a breathtaking cascade.

A catalyst supreme, it sparks the flame,
Accelerating reactions, without any shame.

In laboratories and factories, progress is found,
Dysprosium, a catalyst that knows no bounds.
   Let us celebrate this element's might,
Dysprosium, a beacon of scientific light.
In its atomic wonders, we find inspiration,
A testament to nature's intricate creation.

THIRTEEN

RENEWABLE ENERGY

Dysprosium, a name that echoes in the mist,
A rare element, on the periodic list.
Magnetic and strong, with a shimmering hue,
In its atomic dance, it reveals something new.
    Within its core, secrets are held,
A magnetism that cannot be quelled.
In alloys and magnets, it finds its domain,
Harnessing forces, a power untamed.
    Oh, Dysprosium, a symbol of control,
In the world of magnets, you take a bold role.
With your unique properties, you captivate,
Unleashing potential, as we innovate.
    Let us marvel at this element's might,
Dysprosium, shining in scientific light.

In its atomic wonders, we find fascination,
A testament to nature's intricate creation.
　From technology to renewable energy,
Dysprosium, you shape our destiny.
In your magnetic embrace, we find solace,
A reminder of nature's boundless grace.

FOURTEEN

HIDDEN GEM

Dysprosium, a hidden gem of the periodic table,
A captivating element, its story weaves a fable.
In its atomic realm, secrets lie untold,
A tale of wonder waiting to unfold.

Within its core, a nucleus so strong,
Dysprosium dances to a different song.
Magnetic in nature, it pulls and it attracts,
A force so powerful, it leaves an impact.

In rare earth minerals, its beauty is found,
A symphony of colors, a sight to astound.
From deep within the Earth, it emerges with grace,
A testament to nature's exquisite embrace.

Oh, Dysprosium, a conductor of light,
Illuminating our world, shining ever so bright.
In lasers and lamps, its radiance is seen,
A beacon of hope, where dreams convene.

Let us celebrate this element unique,
Dysprosium, a treasure that we seek.
In its atomic wonders, we find inspiration,
A symbol of possibilities, a source of fascination.

Across the vast expanse, from sky to sea,
Dysprosium, you guide us to what can be.
With your presence, the future takes flight,
A testament to the power of scientific might.

FIFTEEN

GLORIOUS SPARK

Dysprosium, a hidden gem in the land,
With properties rare, like a secret grand.
In the periodic table, you claim your space,
A puzzle piece in nature's embrace.

Oh, Dysprosium, a magnet's delight,
With magnetic strength, you hold things tight.
In motors and generators, your power unfurls,
Driving the world with magnetic whirls.

A beacon of stability, you remain,
Resisting corrosion, never in vain.
In alloys and steel, your strength is renowned,
Forging a path where resilience is found.

Let us celebrate this element's lore,
Dysprosium, a symbol to explore.

In its atomic wonders, we find intrigue,
A testament to elements' mystique.
 From ancient times to modern advances,
Dysprosium, your presence enhances.
In the realm of science, you make your stand,
A remarkable element, oh, so grand.
 Oh, Dysprosium, you captivate our sight,
A chemical marvel, shining with light.
In the realm of elements, you leave your mark,
A testament to nature's glorious spark.

SIXTEEN

DISCOVERY AND EXPLORATION

Dysprosium, an element rare,
With properties beyond compare.
In the depths of the periodic table's array,
You hold secrets that come to light today.

Oh, Dysprosium, magnetic and strong,
In magnets and data storage, you belong.
Your presence alters the course of fate,
Guiding electrons to dance and oscillate.

A conductor of light, you enchant the scene,
In lasers and phosphors, a radiant gleam.
With vibrant hues, you paint the sky,
A cosmic palette, captivating the eye.

Let us celebrate this element's reign,
Dysprosium, an enigma we can't explain.

In its atomic wonders, we find awe,
A testament to nature's mystical law.
   From labs to industries, your influence spreads,
Unleashing innovation from scientific heads.
Oh, Dysprosium, you shape our world's design,
A catalyst for progress, forever divine.
   In the realm of elements, you shine bright,
Dysprosium, a beacon of scientific light.
In your presence, we find inspiration,
A catalyst for discovery and exploration.

SEVENTEEN

ELEMENT EXTRAORDINARY

Dysprosium, element of intrigue,
In the periodic table, you find your league.
With atomic number sixty-six,
Your magnetic properties leave us transfixed.
    Oh, Dysprosium, a conductor of energy,
In magnets and electronics, you hold the key.
With your high magnetic strength,
Efficiency and power, you extend the length.
    A rare earth metal, you stand alone,
In alloys and lasers, your presence is known.
With luminescent hues, you light the way,
A celestial dance in night's dark array.
    Let us celebrate this element's might,
Dysprosium, shining with scientific light.

In its atomic wonders, we find fascination,
A symbol of innovation and exploration.
    From modern technology to sustainable dreams,
Dysprosium, you guide us through the streams.
A catalyst for progress, a symbol of hope,
In your elemental realm, we learn and cope.
    Oh, Dysprosium, element extraordinary,
In your magnetic allure, we find sanctuary.
In labs and industries, your power prevails,
A testament to nature's intricate details.

EIGHTEEN

ATOMIC WONDERS

From the depths of the periodic table's realm,
Dysprosium emerges, a gem at the helm.
With its atomic number, sixty-six,
This element brings wonders that surely transfix.
    Oh, Dysprosium, rare earth metal divine,
In magnets and lasers, your virtues align.
A magnetic force, so strong and untamed,
In motors and generators, your power is famed.
    Let us celebrate this element's grace,
Dysprosium, a marvel of the chemical space.
In its atomic wonders, we find intrigue,
A symbol of progress, a scientific league.
    From advanced technologies to renewable might,
Dysprosium, you shine with radiant light.

A catalyst for change, a catalyst for awe,
In your presence, innovation takes its first draw.
    Oh, Dysprosium, element of intricate design,
Your properties leave us in wonder, so refined.
In laboratories and industries, you play your part,
A testament to nature's scientific art.

NINETEEN

SCIENCE AND INDUSTRY

In the realm of elements, a treasure we find,
Dysprosium, a gem of a different kind.
With atomic elegance, you grace the stage,
A symbol of balance, a chemical sage.

Oh, Dysprosium, your presence sublime,
In magnets and alloys, you mark the time.
With magnetic might, you pull and repel,
A captivating force, where mysteries dwell.

Let us celebrate this element rare,
Dysprosium, with properties beyond compare.
In its atomic wonders, we find delight,
A testament to nature's ingenious might.

From wind turbines to electric cars,
Dysprosium, you drive progress far.

A catalyst for innovation, a catalyst for change,
In your elemental realm, new frontiers we arrange.
   Oh, Dysprosium, element of allure,
In science and industry, you endure.
A symbol of resilience, of strength untold,
In your atomic realm, secrets unfold.

TWENTY

SCIENTIFIC SUCCESS

Oh, Dysprosium, element of strength and might,
In the realm of chemistry, you shine so bright.
With magnetic allure that captivates the soul,
You leave an indelible mark, a fascinating role.

In rare earth metals, you proudly stand,
A beacon of stability, a force in demand.
From magnets to lasers, your presence is felt,
A catalyst for progress, where wonders are dealt.

Let us celebrate your atomic charms,
Dysprosium, the element that disarms.
In laboratories and industries, your potential unfolds,
Unveiling new horizons, where innovation molds.

Oh, Dysprosium, your properties unique,
In science's embrace, you reach the peak.

A symbol of exploration, of the unknown,
In your atomic structure, greatness is sown.
 From sustainable energy to technological feats,
Dysprosium, your impact never retreats.
A testament to nature's intricate design,
You inspire us to push boundaries, to redefine.
 Oh, Dysprosium, element of awe and might,
In the world of chemistry, you shine so bright.
May your legacy endure, a symbol of progress,
Forever celebrated, in scientific success.

TWENTY-ONE

INSPIRE BRILLIANCE

Dysprosium, element of mystique,
In the realm of chemistry, you speak.
With atomic artistry, you dance,
A symphony of possibilities, enhanced.

Oh, Dysprosium, your magnetic charm,
In magnets and alloys, you perform.
A force to reckon, unyielding and strong,
Guiding innovation, where dreams belong.

From wind turbines to electric cars,
Dysprosium, you reach for the stars.
A catalyst for progress, a beacon of light,
In your elemental dance, we take flight.

Oh, Dysprosium, symbol of resilience,
In laboratories, you inspire brilliance.
Unlocking secrets with each discovery,
A testament to human ingenuity.

With luminescent hues, you paint the way,
Illuminating pathways to a brighter day.
Oh, Dysprosium, element of intrigue,
In science's embrace, you continue to intrigue.

May your legacy endure, forever renowned,
Dysprosium, in our hearts, you are crowned.
A symbol of innovation, a source of pride,
In the realm of elements, you will forever reside.

TWENTY-TWO

METAL OF RARE MIGHT

Dysprosium, a metal of rare might,
In the realm of elements, a captivating sight.
With magnetic allure, you draw us near,
A symphony of atoms, harmonious and clear.

Oh, Dysprosium, your presence profound,
In magnets and alloys, your power is renowned.
A spark of innovation, a catalyst for change,
In scientific realms, you rearrange.

From wind turbines to electric cars,
Dysprosium, you drive progress far.
With strength and resilience, you pave the way,
A symbol of ingenuity, day by day.

Oh, Dysprosium, element so sublime,
In laboratories, your secrets we unwind.

You grace us with luminescent hues,
A cosmic dance, where wonders fuse.
    May your legacy endure, steadfast and strong,
Dysprosium, in our world, you belong.
A symbol of possibilities, of untapped might,
In the tapestry of elements, you shine bright.

TWENTY-THREE

CAPTIVATING GRACE

Dysprosium, element of captivating grace,
In the realm of science, you find your place.
With magnetic allure, you pull and entice,
A catalyst for progress, a symbol of might.

Oh, Dysprosium, your atomic dance,
In magnets and lasers, you enhance.
From data storage to medical scans,
You revolutionize with your magnetic hands.

In laboratories, your secrets unfold,
Unveiling mysteries, untold and bold.
A symbol of ingenuity, you ignite,
The flames of discovery, burning bright.

Oh, Dysprosium, element sublime,
Innovations sparked, with each passing time.
In sustainable energy, you take the lead,
Guiding us towards a greener creed.

May your legacy endure, forever strong,
Dysprosium, in the scientific throng.
A symbol of progress, of human endeavor,
In the periodic table, you reign forever.

TWENTY-FOUR

INFINITE WORTH

Dysprosium, an element rare,
With magnetic allure beyond compare.
In the realm of science, you hold your ground,
A symbol of power, profound and renowned.

Oh, Dysprosium, your presence serene,
In magnets and lasers, you intervene.
With strength and resilience, you pave the way,
A catalyst for progress, day by day.

In wind turbines, you spin with grace,
Harnessing energy, in every space.
From green technologies to electronic dreams,
Dysprosium, you illuminate our schemes.

Oh, Dysprosium, element divine,
In laboratories, your wonders align.
Unlocking secrets, expanding our view,
A testament to the curious few.

May your legacy endure, forever bright,
Dysprosium, a guiding light.
A symbol of innovation, of infinite worth,
In the tapestry of elements, you give birth.

TWENTY-FIVE

SUSTAINABLE WAYS

Dysprosium, element rare and true,
In the realm of chemistry, we turn to you.
With magnetic prowess, you astound,
A marvel of nature, profound.

In the core of atoms, your secrets lie,
Unveiling wonders as time passes by.
From alloyed blades to electric might,
Dysprosium, you shine so bright.

Oh, Dysprosium, symbol of innovation,
In laboratories, you spark fascination.
A catalyst for progress, a force to be reckoned,
Guiding us forward, never second.

In wind turbines, you dance with the breeze,
Harnessing energy, with effortless ease.

From green technologies to sustainable ways,
Dysprosium, you shape our modern days.
    May your legacy endure, eternally,
Dysprosium, an element of mystery.
A symbol of resilience, of strength untamed,
In the periodic table, your brilliance is acclaimed.

TWENTY-SIX

SCIENTIFIC THRONG

Dysprosium, element of dreams,
In science's realm, your power gleams.
From rare earth mines, you emerge,
A symbol of progress, a catalyst to surge.
 Oh, Dysprosium, your magnetic might,
Captivates minds, ignites the light.
In data storage and lasers, you excel,
Unleashing possibilities, breaking the spell.
 In laboratories, you reveal the way,
Unveiling secrets, day by day.
A conductor of energy, a conductor of change,
Dysprosium, you rearrange.
 Oh, Dysprosium, so full of grace,
In wind turbines, you find your place.

Harnessing the wind, spinning with might,
Empowering the world, day and night.
　May your legacy endure, forever strong,
Dysprosium, in the scientific throng.
A symbol of resilience, of innovation untold,
In the tapestry of elements, you unfold.

TWENTY-SEVEN

ELECTRIC CARS

Dysprosium, element of rare worth,
In the realm of science, you give birth.
With magnetic allure, you captivate,
A catalyst for progress, we celebrate.

In lasers and lighting, your brilliance shines,
Illuminating paths through complex designs.
A conductor of energy, strong and true,
Dysprosium, we look to you.

Oh, Dysprosium, symbol of innovation,
In laboratories, you spark creation.
Unlocking mysteries, expanding our view,
With each discovery, we marvel anew.

From wind turbines to electric cars,
You power the future, reaching for the stars.

A symbol of resilience, a testament to might,
Dysprosium, you guide us towards the light.

    May your legacy endure, forever renowned,
Dysprosium, in our hearts, you are crowned.
A symbol of progress, of dreams taking flight,
In the tapestry of elements, you shine bright.

TWENTY-EIGHT

ELEMENT DIVINE

Dysprosium, element of rare delight,
In the realm of science, you take flight.
With magnetic allure and atomic grace,
You weave wonders in this cosmic space.

    Oh, Dysprosium, conductor of energy,
In magnets and lasers, you find synergy.
From data storage to medical aid,
Your prowess shines, never to fade.

    In laboratories, your mysteries unfold,
Unveiling secrets, precious and bold.
A symbol of innovation, you lead the way,
Guiding us to a brighter day.

    Oh, Dysprosium, your power astounds,
In wind turbines, your presence resounds.

Harvesting the breeze, with strength untamed,
You bring sustainable dreams, forever untamed.
   May your legacy endure, steadfast and strong,
Dysprosium, in our world, you belong.
A symbol of progress, of infinite worth,
In the periodic table, you claim your berth.
   So, let us celebrate this element divine,
Dysprosium, in our hearts you'll forever shine.
A catalyst for change, a beacon of light,
In the tapestry of elements, your brilliance takes flight.

TWENTY-NINE

FIRMLY STAND

Dysprosium, element of grace,
Within your atoms, mysteries embrace.
A magnetic force, both strong and pure,
You captivate our hearts, that's for sure.

    In lasers and lights, your radiance gleams,
A symphony of colors, like vivid dreams.
Guiding us forward, through the darkest night,
With your luminescence, shining so bright.

    Oh, Dysprosium, conductor of power,
In electric motors, you make things tower.
Efficient and steadfast, you drive the wheels,
With energy harnessed, the world reveals.

    In laboratories, your secrets unfold,
Unveiling wonders, yet to be told.

A symbol of discovery, pushing the boundaries,
Dysprosium, you inspire new foundries.
    May your legacy endure, through the ages,
Dysprosium, sparking scientific stages.
A symbol of progress, of innovation grand,
In the tapestry of elements, you firmly stand.

THIRTY

FULFILLING DREAMS

Dysprosium, a rare and precious find,
In the realm of elements, you're one of a kind.
With atomic grace, you stand tall and proud,
A symbol of strength, in silence, unbowed.

Your magnetic allure, a captivating force,
Guiding compass needles along their course.
In high-tech gadgets and memory chips,
Dysprosium, you bring data to our fingertips.

In the laboratory's gleaming domain,
Scientists embrace you, eager to gain
Insights and knowledge, unraveling the unknown,
With your atomic secrets, they have grown.

Oh, Dysprosium, your luminescence gleams,
Illuminating paths, fulfilling dreams.
A conductor of energy, you empower the world,
From green technology to flags unfurled.

May your legacy endure, forever enshrined,
Dysprosium, a treasure of the mind.
A testament to progress, innovation, and might,
In the periodic table, you shine so bright.

THIRTY-ONE

CATALYST FOR PROGRESS

Dysprosium, element of strength and might,
In the realm of magnets, you take flight.
Aligned with the forces of north and south,
You bring stability, unwavering in your mouth.

A guardian of memories, a keeper of time,
In hard drives and data, your presence does chime.
Preserving moments, stories, and more,
Dysprosium, you safeguard what we adore.

Oh, Dysprosium, conductor of light,
In lasers and phosphors, you shine so bright.
Illuminating the world with a vibrant glow,
From screens to signs, your brilliance does show.

May your legacy endure, steadfast and true,
Dysprosium, we honor the wonders you do.

A symbol of endurance, resilience, and grace,
In the tapestry of elements, you hold a special place.
So let us cherish your atomic embrace,
Dysprosium, with you, we find solace and grace.
A catalyst for progress, a beacon of hope,
In the world of science, you help us cope.

THIRTY-TWO

FASCINATION

In the realm of elements, a jewel so rare,
Dysprosium, with an aura beyond compare.
Your magnetic power, a force to behold,
Drawing us closer, as the stories unfold.

Oh, Dysprosium, conductor of dreams,
In motors and generators, your presence beams.
Unleashing energy with unwavering might,
You propel progress, igniting the light.

A guardian of secrets, locked within your core,
Scientists seek wisdom, forever craving more.
In laboratories, they unravel your mysteries,
Expanding horizons with each new discovery.

May your legacy endure, eternal and strong,
Dysprosium, a symbol of resilience along.

A catalyst of change, in the world we reside,
In the tapestry of elements, you gracefully glide.
    So let us celebrate your atomic embrace,
Dysprosium, a marvel in time and space.
A testament to innovation and exploration,
In the universe of elements, our eternal fascination.

THIRTY-THREE

GUARDIAN OF STABILITY

In the depths of the periodic table's array,
Dysprosium resides, a gem in its own way.
With its magnetic might and atomic charm,
It enchants the world, leaving us in awe and alarm.

Oh, Dysprosium, conductor of forces untamed,
In magnets and alloys, your magic is proclaimed.
You shape the realms of technology and beyond,
Guiding innovation, like a virtuoso's wand.

From wind turbines spinning in the breeze,
To nuclear reactors harnessing the seas,
Dysprosium, you fuel our energy dreams,
With power and efficiency, bursting at the seams.

May your legacy endure, steadfast and true,
Dysprosium, a symbol of progress we pursue.

A guardian of stability, a beacon of light,
In the grand tapestry of elements, shining so bright.
   So let us honor your presence, rare and sublime,
Dysprosium, a catalyst for the passage of time.
With each discovery, we unravel the unknown,
In your atomic embrace, we find a future yet to be shown.

THIRTY-FOUR

DANCE OF ELECTRONS

In the depths of the periodic table's domain,
Dysprosium resides, a treasure to ascertain.
With magnetic allure, you captivate our gaze,
A dance of electrons, in a mesmerizing haze.

Oh, Dysprosium, conductor of control,
In magnets and alloys, you play a vital role.
Manipulating forces, bending them to will,
In motors and turbines, your power instills.

From green technology to futuristic dreams,
Dysprosium, you pave the way, it seems.
A symbol of progress, of precision and might,
In the realm of elements, you shine so bright.

May your legacy endure, unyielding and strong,
Dysprosium, a foundation we rely upon.

A guardian of stability, a guiding star,
In the world of science, you've traveled far.
   So let us celebrate your atomic embrace,
Dysprosium, a marvel of time and space.
With each new discovery, we unlock the door,
To a future where possibilities soar.

ABOUT THE AUTHOR

Walter the Educator is one of the pseudonyms for Walter Anderson. Formally educated in Chemistry, Business, and Education, he is an educator, an author, a diverse entrepreneur, and he is the son of a disabled war veteran. "Walter the Educator" shares his time between educating and creating. He holds interests and owns several creative projects that entertain, enlighten, enhance, and educate, hoping to inspire and motivate you.

Follow, find new works, and stay up to date
with Walter the Educator™
at WaltertheEducator.com

www.ingramcontent.com/pod-product-compliance
Lightning Source LLC
LaVergne TN
LVHW052000060526
838201LV00059B/3762